ALL ABOUT ANIMALS
BIRDS

CHELSEA HOUSE
PUBLISHERS
A Haights Cross Communications Company ®
Philadelphia

First hardcover
library edition published
in the United States of
America in 2005 by Chelsea
House Publishers, a subsidiary
of Haights Cross
Communications.
All rights reserved.

A Haights Cross Communications Company ®

www.chelseahouse.com

Library of Congress Cataloging-in-Publication Data
ISBN: 0-7910-8688-7

Copyright © McRae Books Srl, 2005

The series *All About Animals* was created and produced by:

McRae Books Srl
Borgo S. Croce, 8, 50122, Florence (Italy)
info@mcraebooks.com
e-mail: mcrae@tin.it
www.mcraebooks.com

Publishers: Anne McRae and Marco Nardi
Text: Anita Ganeri
Illustrations: Alessandro Baldanzi, Fiammetta Dogi, Paula
Holguín, Sabrina Marconi, Paola Ravaglia, Studio Stalio
(Ivan Stalio, Alessandro Cantucci, Fabiano Fabbrucci,
Margherita Salvadori)
Graphic Design: Marco Nardi, Yotto Furuya
Layouts: Rebecca Milner, Laura Ottina
Research: Chris Hawkes, Claire Moore, Laura Ottina
Editing: Chris Hawkes, Claire Moore
Color separations: Litocolor, Florence (Italy)
Cutouts: Alman Graphic Design

Printed and bound in China

Acknowledgments
All efforts have been made to obtain and
provide compensation for the copyright to the
photos and artworks in this book in
accordance with legal provisions. Persons who
may nevertheless still have claims are requested
to contact the publishers.

t = top; tl = top left; tr = top right; tc = top center; c = center; cl = center left; cr
= center right; b = bottom; bl = bottom left; br = bottom right; bc = bottom center

The publishers would like to thank the following sources for
their kind permission to reproduce the photos in this book:

Giuliano Cappelli, Florence: 20br, 22br; Contrasto/Corbis: 9c,
33bl; Frederic/Jacana–Overseas: 26c; Lonely Planet Images:
Jason Edwards/LPI 16b, Paul Beinssen/LPI 26c, 27; Marco
Nardi: 32c; Panda Photo, Rome: V.Bretaguolle/Panda 24br, E.
Coppola/Panda 25b; Scala Group, Florence: 32tr; The Image
Works: ©Topham/The Image Works 9bl, 17t, ©Frozen
Images/The Image Works 19, ©Thomas Spangler/The Image
Works 31tl, ©Syracuse Newspapers/Dick Blume/The Image
Works 34bl, ©Michael J. Doolittle/The Image Works 37br.

p. 10: extract from 'The Jay and the Peacock' by Aesop, fable
author (620–560 B.C.); p. 10: extract from 'Nichomachean
Ethics' by Aristotle (350 B.C.); p. 16: extract from 'The Marriage
of Heaven and Hell' by William Blake (1790); p. 19: extract
from 'The Marriage of Heaven and Hell' by William Blake
(1790); p. 20: extract from 'The Recluse: Home at Grasmere' by
William Wordsworth (1798); p. 22: extract from 'The Blue Lion
and Other Essays' by Robert Lynd (1923); p. 29: quotation by
Emily Dickinson (1830–1886); p. 32: quotation by Charles
Lindbergh in an interview shortly before his death in 1974
(1902–1974); p. 35: extract from 'Walden' by Henry David
Thoreau (1854); p. 37: extract from 'Birds and Poets' by John
Burroughs (1887).

ALL ABOUT ANIMALS
BIRDS

Anita Ganeri

Contents

Feathers

Feathers are used for flight, insulation, and streamlining. Colorful feathers help birds identify each other and attract a mate. A large bird like a swan may have 25,000 feathers; a small bird like a hummingbird has about 1,000.

Bird behavior

Many birds behave in ways that help them to survive in their particular habitat. Some birds, such as African honeyguides (right), act in partnership with other animals.

Honeyguides feed on bees' wax. They use calls to guide honey badgers to a bees' nest. The honey badger breaks open the nest so that the honeyguide can get to the wax inside.

What Makes a Bird a Bird?

There are about 9,000 species of birds living all over the world. Birds belong to the Class *Aves* within the animal kingdom. Birds are warm-blooded vertebrates. They lay hard-shelled eggs and are the only animals in the world whose bodies are covered in feathers. Birds have beaks but no teeth. All birds have wings, although some, such as emus and ostriches, cannot fly.

Flying

All birds have wings and most of them can fly. A bird needs to be as light as possible in order to fly well. Flying birds have hollow bones, resulting in extremely lightweight skeletons. A barn owl's (left) wings are fringed with feathers to muffle the sound of its wingbeats as it flies after prey.

Adapting to life

Birds range in size from tiny hummingbirds to huge wandering albatrosses. Each species of bird has its own features that help it survive. These include different shapes of beak and feet, and different colors of feathers for display and camouflage.

Eggs

All birds lay hard-shelled eggs. The shape and color of the eggs varies from species to species. This gull's egg (left) is laid on the ground. Its speckled color helps hide it from predators.

The jacana, or "lily-trotter," has very long toes and claws. These spread its weight and stop it from sinking as it walks across waterlily pads looking for food.

Different diets

Seeds, fruit, worms, insects, nectar, fish, and grass are the most common types of food eaten by birds. Birds' diets vary depending on where they live. Woodpeckers (right) feed on insects that live under the bark of trees.

Birds and humans

The relationship between people and birds dates back thousands of years. People have hunted birds for food and sport, kept them as pets, and worshipped them as gods. Early sailors used birds as guides to find land, while farmers based their weather forecasts on the behavior of birds.

This illustration (left) shows an ancient painting of a giant bird, probably a vulture.

Song and display

Birds have various ways of communicating with each other. They use songs and calls to send warnings, defend their territory, and find a mate. They also use visual signals to send messages. During courtship, many male birds, like this great egret (right), display their feathers to attract a mate.

Protecting birds

There are many conservation groups dedicated to protecting birds. Measures include setting aside undisturbed areas of natural habitat, protecting nesting sites, and establishing captive breeding programs for endangered species. Here (right) lesser white-fronted geese are being taught how to migrate.

Male roadrunners catch lizards and other tasty treats to prove their strength and stamina to females.

Birds on the brink

Today the welfare of birds around the world is under threat. Habitat destruction, pollution, hunting, the pet trade, and man-made disasters, such as oil spills, have all taken their toll. This cormorant (below) has become a victim of an oil spill.

The importance of birds

Birds play a vital part in the natural world. They are part of many food chains, both as predators and prey. Birds that feed on nectar help pollinate plants. Fruit-eating birds help spread seeds in their droppings.

A hummingbird (left) drinking nectar from a rain forest flower.

Courtship and mating

During the breeding season, competition to find a suitable mate is fierce. Male birds have many ways of attracting a female. Some sing or dance. Some show off their plumage. Others, like the roadrunner (above), bring the female gifts of food.

The Evolution of Birds

For millions of years, insects were the only flying animals on Earth. From time to time, small lizard-like creatures took to gliding among the trees on wings formed from flaps of skin. By about 180 million years ago, however, a new group of flying reptiles, the pterosaurs, dominated the skies. They flew by flapping their wings. Meanwhile, the ancestors of modern birds were also evolving. When the dinosaurs and flying reptiles died out 65 million years ago, birds took over the skies.

Archaeopteryx (right) was about the size of a crow. It lived about 150 million years ago.

◑ From gliding to flying

One of the first reptiles to take to the air was Icarosaurus (above). It lived about 220 million years ago. Its wings were formed from flaps of skin stretched over very long ribs. It could not flap its wings to fly, but it could glide long distances.

Laurasia

TETHYS SEA

Gondwanaland

Large, flightless birds are still found in Africa, South America, Australia, and New Zealand. Their evolution matches the break up of the ancient southern supercontinent, Gondwanaland (above).

◑ Archaeopteryx

Most scientists now think that birds evolved from small dinosaurs. In 1861, fossils of the earliest known bird were found in Germany. It was named Archaeopteryx which means "ancient wings." It had claws and a bony tail like its reptile ancestors, but it also had feathers like a modern bird.

Rhamphorhynchus (left) lived about 175 million years ago. It flew along the coasts of southern Europe, hunting for fish.

Archaeopteryx fossil (left).

⊂ Flying reptiles

The name *pterosaur* means "wing-reptile." These reptiles had long, curving wings formed from thick flaps of leathery skin, supported by the elongated fourth fingers of their hands. Powerful muscles connected the arms to the breastbone. These muscles flapped the pterosaurs' wings. Early pterosaurs, like this Rhamphorhynchus, had long tails for balance and steering.

➲ Sea birds

Fossil discoveries have revealed two sea birds — Ichthyornis (left) and Hesperornis (right). Ichthyornis looked like a modern-day tern or gull. It flew over the sea in search of prey. Hesperornis did not have muscles strong enough for flight, but spent its life in the sea, diving for fish.

➲ Other flying animals

While a number of vertebrate animals, such as lizards and snakes, can glide long distances, the only ones capable of true flapping flight are birds and bats. Scientists do not know exactly how bats evolved. It is thought that they are descended from gliding insect-eating creatures that lived in trees and whose arms developed into wings.

➲ Ancient and modern

The ancestors of modern birds date from about 100–65 million years ago. Presbyornis looked like a long-legged duck. It fed on plants in shallow, freshwater lakes. The first big split into many different orders and families of birds occurred about 50 million years ago.

The dodo (below) was about the size of a swan with a large head and curly tail feathers. It could not fly to escape predators.

➲ Large and flightless birds

Some prehistoric birds were too large to fly. Among them was the gigantic Diatryma. A flightless meat-eater, it looked like a modern emu, stood 6 feet, 7 inches (2 m) tall, and had a massive beak. It lived about 60 million years ago in North America.

Moas stood up to 13 feet, 1 inch (4 m) tall. They are the tallest birds ever known.

◖ Extinct birds

The most famous example of an extinct bird species is the dodo. Until the late 17th century, the dodo lived on the Indian island of Mauritius. Within 50 years of the arrival of Dutch sailors, the dodo was extinct, due to excessive hunting and the introduction of predators by the settlers.

➲ The moa

The moa was a large, flightless bird that lived for millions of years in the ancient forests of New Zealand. Until the arrival of people about a thousand years ago, these huge birds had few natural enemies, but hunting, together with a low breeding rate, led to their eventual extinction.

Hunting birds

Vultures, falcons, hawks, eagles, and secretary birds belong to the order *Falconiformes*, or birds of prey. They all share certain characteristics for finding, catching, and killing their prey. These features include sharp eyesight, strong legs and feet, sharp, curved claws, and hooked beaks. Most birds of prey also feed on the remains of dead animals. In the case of vultures, this is their main source of food. Owls belong to a separate order (*Strigiformes*), but are equipped with many similar features, such as sharp eyesight, which help them hunt for prey at night.

Buzzard

Steller's sea eagle

Egyptian vulture

Lanner falcon

Hawk

Great gray owl

Secretary bird

A World of Birds

The ability to fly, combined with warm-blooded bodies, has enabled birds to take advantage of a wide variety of habitats. Birds are found all over the world, from the frozen Arctic and Antarctic to tropical rainforests and barren deserts. There are more than 9,000 species of living birds and they are divided into 23 orders, or groups.

Flightless birds

Some birds, such as penguins, rheas, and ostriches, have wings but cannot fly. Penguins use their wings as flippers for swimming underwater. Ostriches are too heavy to fly, but can run at more than 43.5 miles per hour (70 km/hr). Even a racehorse cannot run that fast!

Penguin

Flamingo

Rhea

Cormorant

Stork

Great blue heron

Ibis

Goose

Blue-footed booby

Pelican

Crowned crane

Pintail duck

Swimming and wading birds

Many birds live on or near fresh water — in ponds, rivers, and lakes — or near the sea. They feed on fish, invertebrates (such as worms and shellfish), and water plants. Many are strong swimmers or have long legs for wading through deep water.

Chicken

☉ Game birds

Game birds, such as turkeys, pheasants, and ptarmigans, belong to the order *Galliformes*. These are stocky birds that nest on the ground and can only fly short distances. Over thousands of years, people have domesticated wild game birds for their meat and eggs.

Turkey

Female argus pheasant

Crossbill finch

Weaver

Arctic ptarmigan

⊃ Perching birds

Over half of all species of living birds belong to the order *Passeriformes*, or perching birds. These are mainly small birds that can be identified by the pattern of their feet. All perching birds have three toes that point forward and one toe that points backward. Finches, weaver birds, swallows, orioles, buntings, starlings, and wrens all belong to this group.

Swallow

Superb blue fairy wren

Starling

Toucan

Swift

Cuckoo

Nightjar

Golden-breasted bunting

Eskimo curlew

Oriole

Ivory-billed woodpecker

Crested ibis

☉ Other common birds

Other common birds that do not fall into any of the other categories include the cuckoo (a bird whose call traditionally signals the start of spring), the woodpecker, the pigeon, the dove (a bird that has become an international symbol of peace), the hummingbird, the parrot, and the swift.

Woodpecker

Spix's macaw

Parrot

Philippine eagle

Kakapo

Californian condor

Hummingbird

Kingfisher

Dove

⊃ Endangered birds

About a fifth of all bird species are currently in danger of extinction and many more are under threat. Hunting, loss of habitat, the use of pesticides, pollution, and the illegal pet trade have put many birds at risk. It is estimated that fewer than 50 adult Californian condors remain in the wild. Human activities have destroyed much of this magnificent bird's natural habitat.

The Bird's Body

There is an amazing variety of different bird species. They range in size from the tiny bee hummingbird, which is no bigger than a bumble bee, to the gigantic ostrich, which stands almost 9 feet, 9 inches (3 m) tall. Although all birds share many similar features, such as feathers, wings, beaks, and two feet, these vary enormously in size, shape, and color.

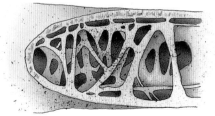

🎧 Skeleton

A bird's skeleton needs to be light, but strong, for flying. A flying bird's bones are hollow to save weight. In many flightless birds, the bones are solid and heavy. Some bones are fused together to give the bird's skeleton extra strength.

Muscles

A flying bird's body is built for flight, with a smooth, streamlined shape. The powerful muscles used to flap the bird's wings are attached to its very broad breastbone.

Eyes and ears

For many birds, sharp eyesight is the most important of all their senses. It is vital for navigation and for finding food. Birds that hunt at night, such as owls, not only have superb vision but also excellent hearing for locating prey.

Hummingbird, long and pointy, for sucking nectar out of flowers.

European bee-eater, long and sharp, for eating insects.

⤵ Beaks

Birds use their beaks to catch and hold food. The size and shape of a bird's beak varies according to what it eats. The illustrations (right) show five different types of beak and what they are used for.

Finch, strong, for cracking seeds.

Falcon, sharp, for tearing flesh.

Pelican, with pouch, for holding fish.

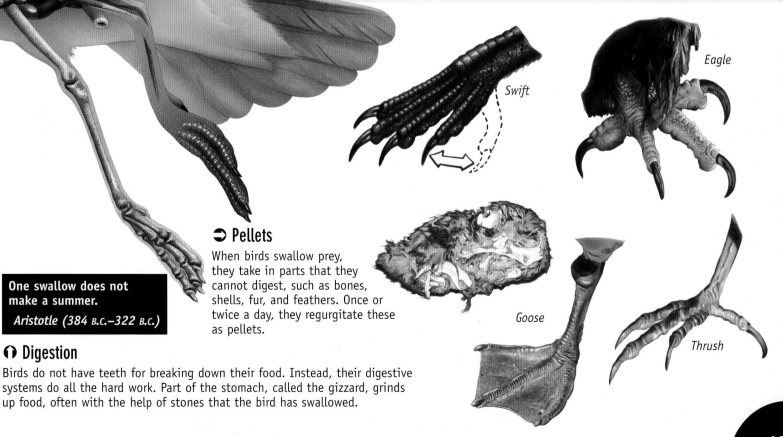

Lungs and air sacks

Birds have very efficient respiratory systems. This enables them to breathe in enough oxygen to release the huge amounts of energy needed for flying. Birds have a large number of air sacks throughout their bodies. These allow them to take in more oxygen with each breath.

Flight feather

➲ Feathers

Like hair and claws in other animals, a bird's feathers are made from a protein called keratin. This makes them extremely strong and flexible. The feathers on a bird's body vary in size and shape, depending on what job they do.

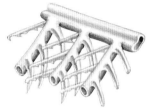

Flight feathers are held together by tiny hooks called "barbules."

Wings

Birds' wings all follow the same pattern, although they vary in size and shape. Long, pointed wings are good for gliding and soaring. Short, stubby wings are good for short flights.

◑ Feet

Birds' feet come in different shapes and sizes, depending on where and how they live. Perching birds, such as thrushes, have three front toes and one rear toe for gripping branches. Birds of prey have sharp, curved claws for catching victims. Many water birds, such as swans and geese, have webbed feet for swimming. Birds that rarely land, such as swifts, have such tiny feet and weak legs that they find walking very difficult.

Swift

Eagle

➲ Pellets

When birds swallow prey, they take in parts that they cannot digest, such as bones, shells, fur, and feathers. Once or twice a day, they regurgitate these as pellets.

Goose

Thrush

One swallow does not make a summer.
Aristotle (384 B.C.–322 B.C.)

◐ Digestion

Birds do not have teeth for breaking down their food. Instead, their digestive systems do all the hard work. Part of the stomach, called the gizzard, grinds up food, often with the help of stones that the bird has swallowed.

Peregrine falcon

Gull

Finch

Buzzard

Different birds have different flight patterns.

How birds fly

Birds owe their flying ability to the shape of their wings. A bird's wings have an airfoil shape, curved on top and flat underneath. As the bird flies, air flows over its wings. The airfoil creates an area of high pressure below the wing and low pressure above it. This pushes the wing up and keeps the birds in the air.

⋂ Speed

The peregrine falcon is the fastest bird alive. In level flight, it can reach speeds of about 62.1 miles per hour (100 km/hr), but in diving down after its prey, it has been known to exceed 174 miles per hour (280 km/hr). It dives with its wings partly folded.

⊃ Hovering

Apart from flapping flight, birds also soar, glide, and hover. The tiny hummingbird (right) is the acrobat of the bird world. By beating its wings an incredible 90 times a second, a hummingbird can hover in front of a flower while it feeds on nectar.

A Life on the Wing

The champions of the air, birds are capable of flying farther and faster than any other animal. Their flying ability has given them a great advantage over other creatures, enabling them to exploit a wide range of food sources and to live in a variety of habitats all over the world. Flying also allows birds to escape from danger.

↻ Gliding

The wandering albatross (below) has the longest wingspan of any bird — over 11 feet, 6 inches (3.5 m) from tip to tip. Albatrosses use their long wings to glide over the ocean, coasting on the wind currents. This form of flying is a good way of saving energy. Albatrosses can glide like this for hours without flapping their wings.

⊆ Take-off

Take-off is the part of flight that uses up most energy. A bird must accelerate quickly to gain enough speed. Small birds simply leap into the air and fly away. Larger birds, such as this gray heron, have to run along the surface of the water to build up enough speed to get airborne.

No bird soars too high if he soars with his own wings.

William Blake (1757–1827)

⮌ Migration

In autumn, many birds leave their summer breeding grounds and fly to warmer climates where food is more plentiful. These journeys are called migrations. Cranes (photo, right) fly in large flocks. They instinctively always follow the same migratory routes to the same places.

In late autumn, Canada geese (left) migrate from northern North America to the southern United States and Mexico.

⮌ Migrating geese

Many migrating birds follow a set formation as they fly. Geese travel in a V-shape, with each bird in the group taking a turn to lead. It is hard work for the lead bird. It has to control the speed and direction of the flock, and keep a lookout for any danger ahead.

↻ Feather care

Birds regularly preen their feathers with their beaks to keep them clean and in good condition. They also coat the feathers in oil to keep them waterproof. The oil comes from a preen gland hidden in the feathers near the bird's tail.

⮌ The swift

Swifts spend most of their lives on the wing, landing briefly only when they are ready to breed. Before this, they may have flown continuously for two to three years. Their long, curved wings allow swifts to fly almost nonstop.

⮌ Amazing journey

The Arctic tern (right) makes the longest migration. In autumn, it leaves its Arctic breeding ground to fly south to Antarctica to feed — a one-way trip of some 12,420 miles (20,000 km). Then, as the southern winter begins, the tern flies north again. In this way, it spends summer at each end of the Earth.

An Arctic tern flies nonstop for about eight months each year and covers over 24,840 miles (40,000 km).

Profile: The Eagle

Eagles are among the largest and most powerful of the world's birds. Among the birds of prey, only condors and some species of vultures are bigger. Soaring high above the ground in search of prey, an eagle is a magnificent sight. No wonder that these awe-inspiring birds have been used to symbolize royalty and power throughout history. Scientists divide eagles into four groups, based on their physical characteristics and behavior. These are fish eagles, snake eagles, harpy eagles, and booted eagles.

Eagle ID

Order: *Falconiformes*
Family: *Accipitridae*
Number of species: 30
Species include: golden eagle (*Aquila chrysaetos*); Indian black eagle (*Ictinaetus*); martial eagle (*Polemaetus bellicosus*); tawny eagle (*Aquila rapax*); Verraux's eagle (*Aquila verreauxii*); wedge-tailed eagle (*Aquila audax*)
Diffusion: all continents except Antarctica
Length: 1 foot, 6 inches–3 feet, 3 inches (45–100 cm)
Prey: large and small mammals, birds, carrion
Number of eggs: 1–7
Incubation: about 34 days

This magnificent eagle feather headdress was worn by a North American warrior. Eagle feathers were thought to possess special powers of strength and bravery, and of healing.

☉ An eagle's body

An eagle's body is designed for hunting. The harpy eagle is one of the most powerful birds of prey in the world. It swoops low over the forest on its long, broad wings, using its fearsome talons to seize prey from the trees. It carries its prey back to its nest and rips it apart with its razor-sharp, curved beak.

↻ Eagle nests

Eagles, like this golden eagle, build huge nests of sticks and twigs, high up in trees or on clifftops. These nests are called eyries. During the nesting season, each pair of eagles defends its nest fiercely. Many return to the same nest year after year.

⊃ Young eagles

Young eagles are called eaglets. Once hatched, the chicks grow quickly. After two to three weeks they begin to grow feathers to replace their grayish down. Soon they are able to tear up their food. The eaglets are always hungry, keeping their parents busy hunting for prey. They leave the nest for the first time when they are about 11–12 weeks old.

↻ Prey

This white-bellied fish eagle circles above the ocean, using its superb eyesight to spot fish below the water. Then it dives down with its razor-sharp talons outstretched, to catch its prey. Eagles also prey on birds and mammals, including those as large as monkeys, lambs, and young deer.

The bald eagle, with its striking white head, belongs to the group of fish eagles. These huge birds mainly eat fish. They follow other fish-eating birds, then swoop low over the water and snatch up the fish. The bald eagle is only found in North America and is the national bird of the United States.

Survival Skills

In order to survive, a bird needs a variety of special skills. It must be able to find enough food and water and defend itself against predators. In the breeding season, it must find a mate and a safe place to build its nest. In general, the larger a bird is, the longer it lives. Most wild birds, however, do not die of old age, but fall victim to man-made and natural hazards.

In Africa, oxpeckers perch on the backs of animals, such as giraffes, rhinos and antelopes. They pick off and eat the ticks and lice that crawl over the animal's skin.

Happier of happy though I be, like them I cannot take possession of the sky, mount with a thoughtless impulse, and wheel there, one of a mighty multitude whose way and motion is a harmony and dance magnificent.

William Wordsworth (1770–1850)

The woodpecker finch of the Galapagos (left) uses a twig or thorn to spear insects from under tree bark. It is one of only a handful of animals that uses a type of tool to find food.

☰ Finding food

Birds need to eat frequently in order to survive. The smaller a bird is, the more time it needs to spend feeding. Some small birds have to feed almost continuously. Large birds may be able to go without food for several days. Birds have many different strategies for finding their food. Some have highly specialized beaks or feet, while some depend on other animals to help them.

☊ Sleeping and resting

Most birds search for food by day and sleep at night. Many also take short naps during the day. At dusk, they head for a roosting site, such as a branch or ledge. Flamingoes rest standing on one leg in the water, with their heads tucked under their wings.

➲ Communicating

Birds use calls and songs to communicate with each other. Baby birds use different calls to tell their parents that they are hungry, injured, or afraid. Chicks seem to recognize and react to their parents' calls even before they hatch.

↺ Scaring predators

The sun bittern's brown and gray plumage makes it difficult to spot among the undergrowth of the South American rain forest. If a predator approaches its nest, however, the bittern puts on a threatening display. It spreads out its wing and tail feathers, showing the dark patches, which glare out like a pair of eyes.

This mother seagull is teaching her chick to make and recognize calls.

➲ Establishing a territory

In spring, the male great reed warbler marks out his territory. It includes a quiet nesting site where the female can lay her eggs. Then the male perches on a tall reed and starts to sing his shrill song to attract a mate.

↺ ∩ Blending in

Year-round camouflage hides the ptarmigan from predators in its Arctic home. In spring, its speckled brown plumage blends in with the tundra ground (left). In winter, its white coat perfectly matches the ice and snow (above).

↻ Working together

Some birds greatly increase their chance of finding food by working together. When a flock of blue-footed boobies locates a school of fish, it circles above it. Then, in response to calls by a few of the birds, the boobies dive and snap up the fish in their beaks.

↺ Courtship rituals

Courtship rituals help some birds identify a suitable mate for breeding. Great crested grebes perform an elaborate courtship dance on the water. The dance starts with the grebes shaking their heads at each other. It ends with them diving for weeds, which they then present to each other.

➲ Showing off

Apart from song and calls, male birds have many other ingenious ways of winning a mate. Some rely on bright colors or displays to attract a female's attention. Male frigate birds tip their heads backwards and inflate their throat sacks like bright red balloons. They also shake their wings and clap their beaks, making a loud gobbling sound.

↻ Adapting to new environments

All over the world, birds' natural habitats are being destroyed. To survive, birds have to adapt to new environments and find new sources of food. In Britain, some tits have learned to pierce aluminum milk bottle tops with their beaks to get at the nourishing milk.

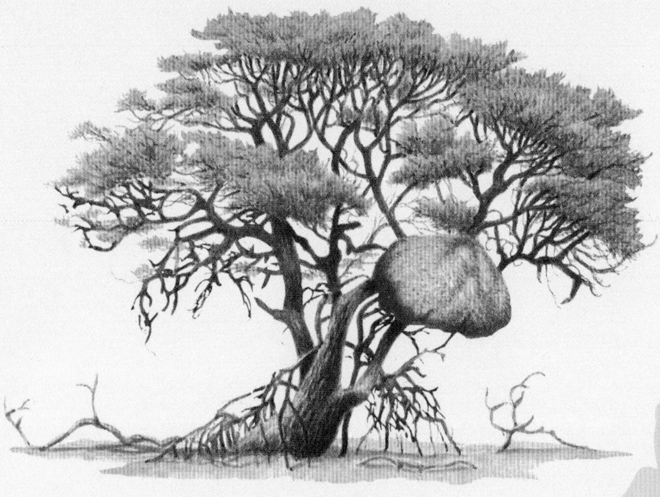

Left: Groups of weaver birds in Africa build huge communal nests in the trees. Each pair of birds has its own living space.

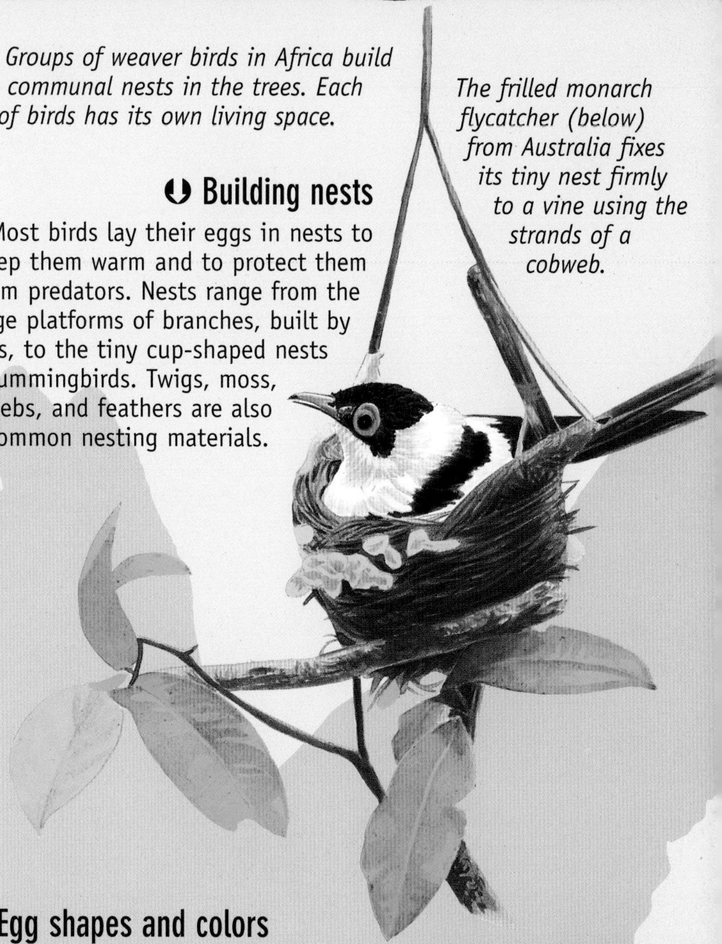

The frilled monarch flycatcher (below) from Australia fixes its tiny nest firmly to a vine using the strands of a cobweb.

Building nests

Most birds lay their eggs in nests to keep them warm and to protect them from predators. Nests range from the huge platforms of branches, built by eagles, to the tiny cup-shaped nests of hummingbirds. Twigs, moss, cobwebs, and feathers are also common nesting materials.

Eggs, Nests, and Hatchlings

Like their reptile ancestors, all female birds lay eggs. The egg provides a supply of food for the growing bird inside, surrounded by a thin, but tough, protective shell. The eggshell has thousands of tiny pores in it, through which air can pass to and fro for the young bird to breathe.

Egg shapes and colors

Birds' eggs come in a wide range of sizes, shapes, and colors. Birds that nest in burrows or holes lay white eggs because they do not need to be camouflaged. Birds that nest in the open lay speckled or dappled eggs to hide them from predators.

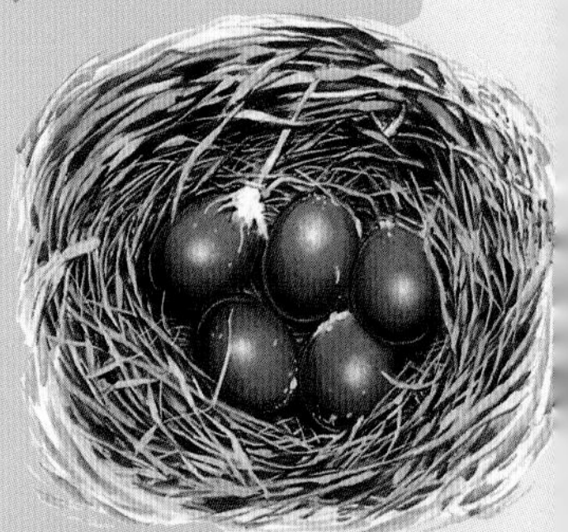

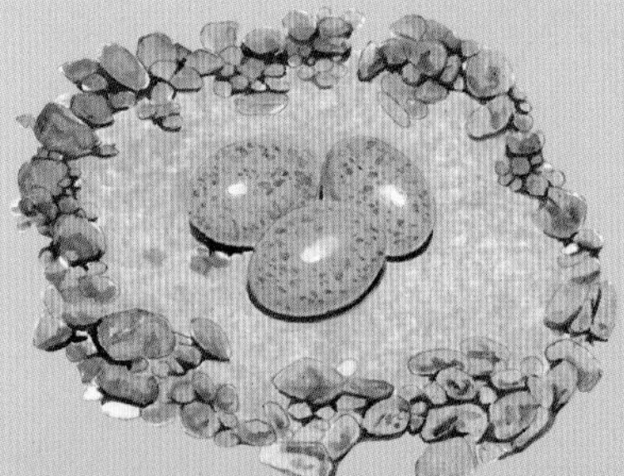

There is nothing in which the birds differ more from man than the way in which they can build and yet leave a landscape as it was before.
Robert Lynd (1892–1970)

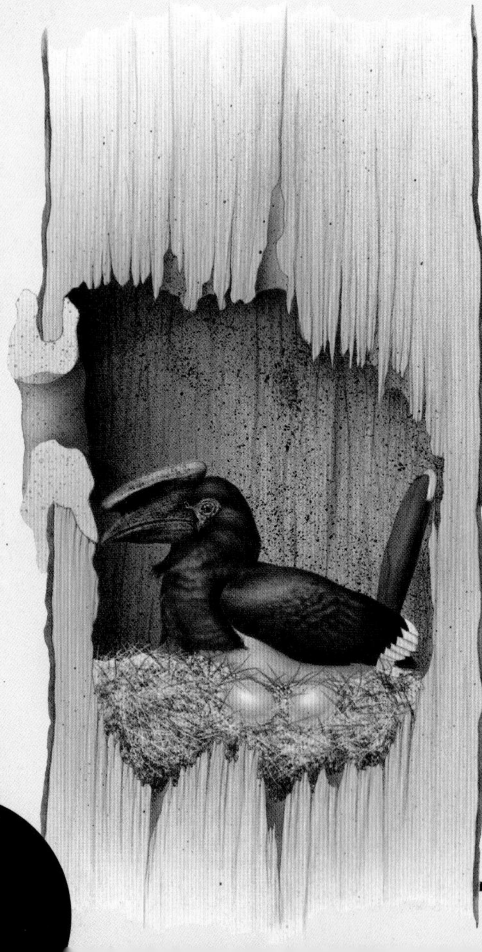

The female hornbill (left) nests inside a hollow tree. Then she seals the entrance with mud and droppings to prevent snakes and other predators from stealing her eggs. The male passes food to her through a tiny hole.

Parental care

Most birds are caring parents. They look after their young until they are old enough to fend for themselves. Swans can be very aggressive if their cygnets are threatened. At the first sign of danger, the cygnets climb on to their mother's back and nestle safely into her feathers. Young ostrich chicks follow their father wherever he goes.

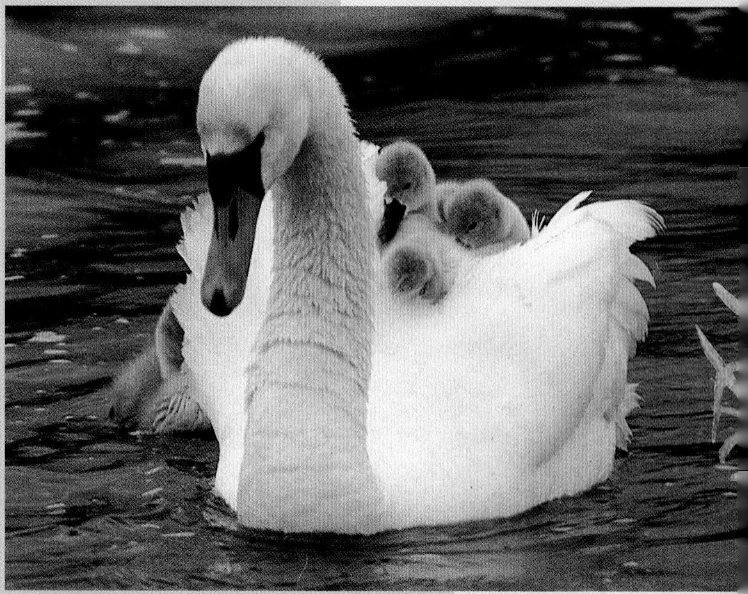

Hatching out

The eggs need to be kept warm so that the chicks inside can develop properly. Usually the female sits on the eggs to keep them warm. This is called the incubation period and lasts from about 10–60 days. Then the chick must break out of its shell, using the bony "egg-tooth" on its beak to chip its way out. Some chicks take less than an hour to hatch. Albatross chicks, though, can take several days.

The female European cuckoo does not build her own nest. Instead, she lays her eggs in another bird's nest. When the cuckoo eggs hatch, the large chick will push the other eggs out of the nest to get its foster parents' full attention.

Hungry chicks

Some birds, such as ducks and geese, hatch in a well-developed state, but many newly hatched chicks cannot see, have no feathers and are helpless. They depend on their parents to bring them food and eat almost constantly. By the time they are ten days old, they may have grown to ten times their original birth weight.

Seagull chicks (below) learn to peck at a red spot on their mothers' beaks to get food. The tapping stimulates the mother to regurgitate food for them.

First flight

Before they can look after themselves, baby birds have to learn to fly. This involves plenty of trial and error before they get it right. Hoatzins live in the rain forests of South America. If the newly hatched chick is threatened, it does not fly but jumps into the water to escape. Then it climbs back up to its nest using its beak and special claws on its wings.

Laying eggs

A female ostrich lays her eggs in a shallow scrape on the ground. She and her mate then take turns to guard and incubate the eggs. Other females also lay their eggs in the nest. These eggs are pushed out by the real nest-owners.

Flightless Birds

Many flightless birds evolved on islands with no natural predators. Because they did not need to escape from danger, the birds lost their ability to fly. Other flightless birds grew too heavy to fly. Instead, they developed other features, such as long legs for running away quickly from enemies.

The kakapo is the only flightless parrot. This bird is vulnerable to introduced predators, such as cats and rats, and is now seriously endangered.

↺ The kiwi

The kiwi is New Zealand's national bird. The size of a chicken, it lives in burrows on the forest floor. A nocturnal bird, it uses its sense of smell to sniff out earthworms, then it pries them from the ground with its long, curved bill.

∩ New Zealand bird life

Before the first humans arrived about 1,000 years ago, the islands of New Zealand had no large predators and we. a paradise for birds, many of which were flightless. Many species became extinct and today only a few are left. They include the kakapo, a nocturnal parrot, and the takahe, a type of rail.

By the mid-19th century, the takahe was thought to have become extinct. Then, in 1948, a small number of birds were rediscovered in a remote region of New Zealand's South Island.

Emperor penguins breed on the Antarctic ice in the middle of winter. The female lays an egg, then leaves the male to incubate it on his feet for the next two months. The males huddle together for warmth (below).

The female kiwi (above) lays one huge egg that weighs about a quarter of her own body weight Then she leaves it for the male to incubate.

⇒ Penguins

With their black and white plumage and waddling walk, penguins are among the most striking of birds. On land, penguins can look clumsy and unsteady on their feet. In the sea, though, they are elegant swimmers, using their wings as flippers to swim at great speed after fish and squid. Their short legs and tails act as rudders.

⊂ Kagu

The flightless kagu lives in small patches of forest on the island of New Caledonia in the Pacific. Once prized for its crest feathers, it is now highly endangered.

⊃ The flightless cormorant

Cormorants reached the Galapagos islands long ago from mainland South America. Once there, they lost their ability to fly because they could find all the food they needed simply by diving into the sea.

⊃ The ostrich

The African ostrich is the world's largest living bird. Males stand 8 feet, 2 inches (2.5 m) tall and can weigh up to 330 pounds (150 kg); females are smaller. Ostriches also lay the largest eggs of any bird. They measure about 8 inches (20 cm) long, with immensely strong shells that could easily support the weight of an average human. Ostriches graze on leaves, shoots, flowers, and seeds. While feeding, they frequently raise their heads to keep a lookout for danger.

The rarely seen cassowary (left) lives deep in the rain forest. It is thought to use the bony helmet on its head to push its way through the undergrowth, and perhaps also to dig for food.

Unlike most birds, the ostrich has only two toes which are highly adapted for fast running. Ostriches can reach over 43.4 miles per hour (70 km/hr) in short bursts.

An emu's wings are tiny and hidden under its shaggy feathers. It uses its long legs instead to walk long distances and to escape from danger.

⊂ Flightless giants

Millions of years ago, huge, flightless birds walked the Earth. Their ancestors still survive today. They include the ostrich in Africa, the rhea in South America, and the emu and cassowary in Australia and New Guinea. The rhea (left) lives on the grasslands of South America. It stands up to 4 feet, 11 inches (1.5 m) high and feeds mainly on plants and insects.

Profile: The Parrot

The family of parrots is called *Psittacidae* and consists of more than 300 species. These beautiful birds are found throughout the tropics, from South America to Australia. Most are tree dwellers. They range in size from tiny budgerigars to huge and spectacular macaws. All parrots have certain features in common. Their beaks are short and strongly hooked. Their feet have two forward-pointing inner toes, while the outer toes point backward to give a very powerful grip.

⊂ Flight

Parrots vary in their flying ability. Generally, small species fly faster than larger species, but macaws, like this ara macaw, are fast fliers despite their size. At another extreme, the kakapo from New Zealand is the only flightless parrot.

⊍ The biggest parrot

Macaws are the largest members of the parrot family. The hyacinth macaw can grow up to 3 feet, 3 inches (1 m) long, including its very long tail. This magnificent bird lives in the forests of Brazil and eastern Bolivia. Deforestation and illegal hunting for the pet trade have pushed it to the brink of extinction.

A parrot's beak is extremely adaptable. It can be used for delicate preening or for crushing the hardest seeds and nuts. It also acts as a third foot for climbing among the trees.

Parrot ID

Order: *Psittaciformes*
Family: *Psittacidae* (parrots)
Number of species: 328
Species include: African grey parrot (*Psittacus erithacus*); blue-and-yellow macaw (*Ara ararauna*); budgerigar (*Melopsittacus undulatus*); rainbow lorikeet (*Trichoglossus haematodus*); sulphur-crested cockatoo (*Cacatua galerita*)
Diffusion: Central and South America, North America, Africa, Southeast Asia, Australia
Length: 3.5–40 inches (9–100 cm)
Diet: fruit, buds, seeds, pollen, and nectar
Number of eggs: 2–8
Incubation period: 17–35 days
Nestling period: 21–70 days

⊂ Mates for life

Many parrots mate for life with the same partner. These masked lovebirds form a very close bond. They almost always stay together and frequently feed and preen each other. Most parrots nest in holes in tree trunks. The female incubates the eggs, but both parents take turns finding food for the chicks.

Rainbow lorikeets may hang upside-down for better camouflage.

⊃ Lories and lorikeets

Lories and lorikeets have longer, narrower beaks than their seed-crushing cousins. They also have tiny brush-like bristles on the tips of their tongues. These adaptations help them collect pollen and nectar, their main source of food.

Parrots are famous for their brightly colored plumage. In fact, most species are green and, as a result, are well camouflaged among the tree tops. Some of the larger species, however, such as the tropical macaws and this eclectus parrot, are among the most brilliantly colored of all birds.

Mythical Birds

From ancient times, birds have appeared in myths and legends from all over the world. Because of their ability to fly — one thing that humans cannot do — they were seen as creatures with magical powers, the messengers and vehicles of the gods, and even as gods themselves.

⟳ The Thunderbird

According to Native American myths, storms are caused by a gigantic eagle, called the Thunderbird. The rumble of thunder is the beating of the Thunderbird's wings. Lightning flashed from its eyes and beak. Despite its awesome power, the Thunderbird is considered a good spirit, bringing welcome rain to the parched Earth.

↻ Egyptian bird gods

In ancient Egypt, many of the gods were associated with birds. The ibis was sacred to Thoth (below), god of the moon and knowledge, who was often shown with an ibis's head. Horus, the protector god of the kings and god of the sky, was depicted with the head of a falcon.

The benu bird (above) was an imaginary bird, resembling a heron. At the temple of the sun god in Heliopolis, Egypt, it was revered as the first deity.

↻ Eagles in religion

Because of its size and magnificence, the eagle has always been viewed with awe and respect. Early people associated it with the sun god, while kings and emperors adopted it as the symbol of their power and might. In Mesopotamian religion, many of the gods were represented as birds, including eagles.

Left: An eagle-headed Mesopotamian god.

⟳ The griffin

The magical griffin was a mixture of two royal animals — the lion and the eagle. Said to be the size of a wolf, it possessed supernatural strength. In legend, it symbolized honor and wisdom, but also evil and pride.

↺ Eagle king

Garuda, the divine bird of Hindu myth, had an eagle's head, wings, and feet. He carried the great god, Vishnu, and his wife, Lakshmi, through the universe.

Phoenix from the ashes

One of the most famous of all mythical birds was the fabulous phoenix. Legend said that, after 500 years, the phoenix sang its last song, then burned itself in its nest, which was made from spices. From the ashes, a new phoenix was born.

The roc

A giant bird from Arabian and Persian myths, the roc was said to feed its young on elephants, which it carried off in its claws. This legend may have been based on the huge, flightless elephant bird from Madagascar.

The head of a feather god from Hawaii.

The simurgh

In Persian myths, the simurgh was a gigantic bird, with a wingspan that could block out the sun, and the head of a dog or man. It lived in the Tree of Knowledge and was able to cure any wound or illness by a simple touch of its wings.

Sacred feathers

In many cultures, birds' feathers are believed to have a sacred meaning or magical powers. In Hawaii, people made images of the gods, covered with feathers. The chiefs carried these on poles as their portable, personal gods.

> I hope you love birds too. It is economical ...
> It saves going to heaven.
> *Emily Dickinson (1830–86)*

Prophetic birds

Some birds are believed to be able to foretell the future. Large, black birds, such as ravens, were often seen as omens of bad luck or death. They were supposed to be able to see into the future and tell if a death was on its way. In Norse myth, the god Odin sent out two ravens every day to bring him news of what was happening in the world.

The raven is an important figure in many Native American myths. Here, the raven finds the first people inside a clamshell.

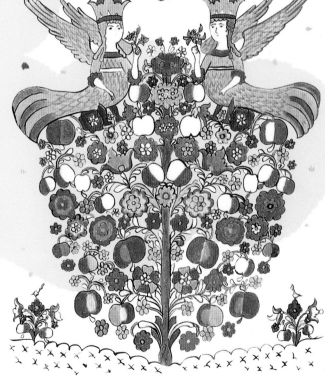

Heavenly birds

Birds are often depicted as the messengers of the gods who fly between heaven and earth. In many religions, angels are associated with birds. The sirin of Slavic folklore was a bird of paradise with the face of a young girl.

Useful Birds

Birds have played an important part in people's lives for thousands of years. Humans have long hunted birds for food and sport, with traps, snares, and, later, guns. The discovery that some types of bird, especially fowl, could be easily domesticated, led to the widespread raising of chickens, turkeys, ducks, and geese for their meat and eggs.

◐ Helpful wild birds

Many types of wild bird are also useful to people. Some birds help farmers by eating weeds, insects, and other pests that damage crops. A few birds are considered pests, such as pigeons and starlings. Their sheer numbers make them a nuisance in many North American and European cities.

◑ Ducks and geese

Geese were among the first birds to be domesticated. Pictures of geese are found in carvings and paintings dating back to the times of the Sumerians and ancient Egyptians. Along with ducks, geese have long been an important food source, especially in eastern Europe and China.

This mosaic from ancient Rome shows a cockfight about to begin. On the table is a bag of betting money. This cruel sport was very popular in the ancient world.

These geese form part of a wall painting from ancient Egypt. Geese and other waterbirds were hunted for food with nets.

◐ Pigeon post

Descended from the wild rock dove, pigeons have been domesticated for centuries. Carrier pigeons are the most closely linked to humans. Pigeons have a strong homing instinct and will fly long distances back to the place where they roost or breed. Before modern communications, people used carrier pigeons to send messages.

⊃ Domestic fowl

The most common domestic bird is the hen or chicken. Its wild ancestors were jungle fowl from Southeast Asia. They were domesticated at least 3,000 years ago. Meat and eggs from chickens are eaten as a staple food all over the world.

◐ Turkeys

Wild turkeys were probably first domesticated in Mexico where they were kept in aviaries. They were brought to Europe in the 16th century and became highly prized for their tasty meat. Today, they are traditionally eaten as part of Christmas or Thanksgiving meals.

A woman feeding her small flock of ducks and chickens.

In China and Japan, another bird, the cormorant, was also trained to catch fish (left). Cormorants are still sometimes used in this way. They are attached to the fisherman's boat by a long leash.

The ancient sport of falconry, or hunting with falcons, may first have been practiced in China thousands of years ago. It remained very popular for centuries. Falcons, hawks, and kestrels were trained to catch rabbits and hares, and then to return to their handler. Falconers wore thick leather gloves to protect their hands from the birds' sharp claws.

This fashionable lady of the early 1900s wears feathers in her hat.

An African shaman with a feather headdress. In some cultures, feathers are thought to have magical properties.

◑ Feathers

All over the world, birds' feathers, particularly those from brightly colored tropical birds, have been used to decorate clothes and costumes. In some societies, they are seen as status symbols or as symbols of royalty. Feathers, especially goose down, have also been used to stuff quilts and pillows.

⊃ Eggs

Birds' eggs are an important source of food for people all over the world. They are also powerful symbols. In the Christian religion, the chocolate eggs exchanged at Easter represent the resurrection of Jesus when he came back to life. The eggs are seen as symbols of new life.

Bird Fancy

Throughout history, birds have provided a source of inspiration to artists, writers, dancers, and musicians. Their beauty, song, and ability to fly have influenced countless works of art, from great paintings and poems, to company logos and postage stamps. Birds are also important symbols — of qualities such as wisdom, freedom, and spirituality.

⇨ Birds in art

From prehistoric times, people have made drawings, carvings, and sculptures of birds. A bird-headed human was one of the earliest figures in art, found in many cultures from ancient Egypt to Polynesia. In 17th- and 18th-century Europe, birds were also a favorite theme. During this time, the New World was discovered. Painting and prints of new and exotic bird species were highly prized.

An illustration (right) by the French-born American naturalist John Audubon (1785–1851), one of the most famous of all bird painters.

A Roman mosaic showing doves drinking water from a golden goblet. It comes from Hadrian's villa near Rome.

> I realized that if I had to choose, I would rather have birds than airplanes.
>
> *Charles Lindbergh (1902–1974)*

A dove on the poster of the 1960s Woodstock rock festival, symbolizing love and peace.

⊂ Symbolic birds

Some birds have become important symbols because of the qualities they are believed to possess. The dove is a universal symbol of peace, love, and faithfulness. In the Christian Bible, Noah released a dove from the Ark. It returned with an olive branch, a sign that the flood was over and of God's forgiveness.

The peacock (left) is the national bird of India, admired for its beautiful tail feathers. It is particularly associated with the god Krishna.

Since ancient times, owls (below) have been symbols of wisdom. In ancient Greece, the owl was the bird of Athena, goddess of wisdom and war.

Cranes (left) have long been regarded as bringers of good fortune. They were symbols of long life because they were believed to live for a thousand years.

⊃ Birds in myth and folklore

Myths and folklore are rich with stories about birds. In European folktales, the white stork was said to bring babies to parents. This legend may have started because storks are caring parents and very protective of their young.

↻ Birds in literature

Literature abounds in bird stories and legends. In Western literature, birds appear in the works of the Greek writers Homer and Herodotus, and later in Dante and Shakespeare. Birds are favorite characters in nursery rhymes and children's stories too. The skylark is often used by poets as a symbol of spiritual enlightenment.

The skylark's musical song has also inspired many works of poetry.

⊂ National birds

Many countries have adopted birds as their national emblems. The quetzal is the national bird of Guatemala. This magnificent bird lives in the rain forests of Central America. The Aztec and Maya worshipped the quetzal as the god of the wind, and prized the male's shimmering green tail feathers for ceremonial costumes.

Guatemala's currency is called the quetzal after its beautiful national bird.

⊃ Bird dances and costumes

In many indigenous cultures, dances based on the movements of birds are performed, among other things, to summon rain, as fertility rituals and to bring a good harvest. Dancers wear elaborate costumes and headdresses, stunningly decorated with birds' feathers.

In African dances, masks (such as the one on the left) represent the spirits of nature. This mask symbolizes an owl, nature's spirit of the night.

An Aboriginal from Australia, dressed in eagle feathers, performing a dance that tells the story of how his ancestors were threatened by man-eating eagles.

Papageno, the bird catcher, from Mozart's opera The Magic Flute.

⊂ Showbiz birds

More recently, birds have starred in films and cartoons. In Alfred Hitchcock's chilling film, *The Birds*, a flock of birds gradually takes over a city and starts to attack its inhabitants. On a lighter note, two of the most popular cartoon birds are the canary, Tweety Pie, and Beep Beep, the roadrunner.

⊃ Birds in music

The musical sound of birdsong has inspired many singers, composers, and musicians. Everything from pop songs to the grandest classical operas has been influenced by the melodic songs of birds, particularly those of birds such as skylarks and nightingales. Many modern pop groups, such as the Eagles, the Housemartins, and Hawkwind, have taken their names from birds.

A sulphur-crested cockatoo.

Pet Birds

Birds have been kept as pets for thousands of years. The most popular breeds include parrots, budgerigars, and cockatoos. Keeping any bird as a pet needs careful thought and preparation. Pet birds require a large cage or an aviary, with the opportunity to fly freely. Birds should only be obtained from a good breeder or animal rescue, and should be captive bred. Many species of bird are now endangered in the wild because of illegal trapping for the pet trade.

◑ Parrots

Large parrots, such as African greys, Amazons, and macaws, can make good pets, but they need plenty of attention and stimulation and will start plucking their feathers out if they get stressed or bored. Given the right conditions, though, they can live happily for 25–30 years, and even longer.

◑ Cockatoos and cockatiels

Many species of cockatoos are kept as pets. Cockatoos have strong personalities. Most are very sociable and need lots of attention. They can be noisy and destructive if they do not get their own way! Cockatiels come from Australia. They are quiet, gentle birds and learn to speak, whistle, and sing.

A pair of cockatiels.

◑ Budgerigars

Budgerigars, or budgies, are among the most popular pet birds. Originally from Australia, these small parrots were introduced to other parts of the world in the mid-19th century. Wild budgies have green plumage, but breeding has produced many other colors, including blue and yellow.

☾ Mynah mimics

Mynah birds are famous for their ability to speak. They can mimic any sound they hear, from ringing phones to babies crying and dogs barking. The Greater Indian Hill mynah (left) is the most popular breed. It is easy to tame and train, especially if you get a young bird of six months old or less. Given a healthy diet, mynahs can live for up to 25 years.

☽ Toucan

Among the most distinctive of birds, the toucan lives in the lush rain forests of South America. Local rain forest people often keep toucans as pets, but allow them to fly freely. Toucans are difficult to keep in captivity and should only be kept as pets by expert bird-owners. These birds need plenty of room and a carefully controlled diet. Only captive-bred toucans should ever be kept as pets.

➲ Canaries

Canaries are a type of finch, native to the Canary Islands after which they are named. These small birds have been kept as pets for hundreds of years. They were also bred to work in coal mines, detecting poisonous gases.

➲ Pekin robin

With its beautiful song, the Pekin robin is a popular pet. It mixes well with other birds and is best kept in a large aviary, with plenty of leafy foliage where it can find insects to eat and places to hide and nest.

In the wild, toucans use their huge bills to reach for food and send signals.

A Gouldian finch.

> I once had a sparrow alight upon my shoulder for a moment ... and I felt that I was more distinguished by that circumstance than I should have been by any epaulet I could have worn.
>
> *Henry David Thoreau (1817–1862)*

A pair of Zebra finches.

☾ Finches

Finches are small birds that are relatively easy to keep as pets. It is best to keep them in an aviary, as part of a group. Popular breeds of finches include the Gouldian, Zebra, Spice, and Society. The Society finch was bred in captivity by breeding several different species of finch together. Unfortunately, many wild finches are caught illegally for the pet trade, causing a great deal of suffering to the birds.

Caring for Birds

Keeping birds as pets is a great responsibility. Captive birds cannot find food for themselves, so it is up to their human owners to provide them with a healthy diet. They also need water, a spacious place to live, and plenty of attention. In general, more unusual birds are harder to care for and should not be kept as pets.

⊃ Bird toys

Birds kept in pairs or groups have each other for company, but if a bird is kept alone, it can easily get bored and start to pull out its feathers or make a lot of noise. Caged birds can benefit from a variety of toys to keep them alert and interested. Bells, cardboard tubes, and chew sticks all make good toys for birds.

⌒ A balanced diet

It is vital that a pet bird has a balanced diet to keep it healthy and happy. In the wild, birds eat a variety of fresh food, including seeds, nuts, fruit, and fish or meat. They must always have fresh drinking water. A vet or breeder will advise on the best diet for different types of birds.

If birds are kept in a small cage inside the house, they should be allowed to fly freely as often as possible.

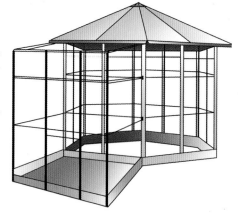

A large aviary provides pet birds with a home that is more like their natural environment.

⊂ Cages and aviaries

Birds are used to flying freely, so pet birds should be given a large cage or an aviary, with plenty of space for flying and exercise. The bigger the cage, the better. There should be perches on different levels and dishes for food and water. The cage should be placed out of drafts and direct sunlight and will need regular cleaning.

Healthcare

Pet birds need a regular veterinary checkup to make sure that they are healthy. Any problems, such as changes in feeding habits or general behavior, could be warning signs of illness or injury. If a bird fluffs up its feathers (as shown below) and does not move for several hours, it should be taken to a vet.

Providing a nesting box is a good way of encouraging garden birds to stay. It should be placed out of direct sunlight and out of the reach of cats.

Bathing

Wild birds like to bathe in mud or water to keep their feathers in good condition. Some pet birds, such as Amazon parrots, enjoy having a shower. The shower spray mimics the falling rain in their natural rain forest home.

Garden birds

During winter, many wild birds struggle to find enough food to stay alive. A good way of attracting birds into the garden is by providing them with a supply of food. Special dispensers can be filled with nuts and seeds. Balls made of fat and seed can be hung from trees.

The very idea of a bird is a symbol and a suggestion to the poet. A bird seems to be at the top of the scale, so vehement and intense his life ... The beautiful vagabonds, endowed with every grace, masters of all climes, and knowing no bounds — how many human aspirations are realised in their free, holiday lives — and how many suggestions to the poet in their flight and song!

John Burroughs (1837–1921)

Birdwatching

Watching wild birds can be a rewarding hobby. You will quickly learn to identify different species by their appearance and calls. A field guide and a pair of binoculars are a must for beginners. Always avoid disturbing the birds, especially if they are nesting.

Index